TUNNEL RATS VERSION 2.0

FIGHTING AND WINNING FUTURE WAR IN A SUBTERRANEAN ENVIRONMENT

BRIAN T. SPILLANE

NIMBLE BOOKS: THE AI LAB FOR BOOK-LOVERS
~ FRED ZIMMERMAN, EDITOR ~

Humans and AI making books richer, more diverse, and more surprising.

Publishing Information

(c) 2023 Nimble Books LLC
ISBN: 978-1-934840-76-4
AI Lab for Book-Lovers
Humans and AI making books richer, more diverse, and more surprising.

AI-generated Keyword Phrases

subterranean warfare; United States military; combatting subterranean threats; historical examples; relevance in the 21st century; lack of focus; lack of doctrine; lack of training; lack of equipment; changes in concepts; changes in capabilities; changes in technologies; detecting underground threats; destroying underground threats; Israeli Defense Force; specialized unit; integration of skill sets; task organized teams; manned, trained, and equipped; subterranean environment; overcoming challenges; achieving success; future conflicts;

Publishing Information 2
AI-generated Keyword Phrases 2
Abstracts 4
    TL;DR (one word) 4
    Explain It To Me Like I'm Five Years Old 4
    TL;DR (vanilla) 4
    Scientific Style 5
    Action Items 5
Viewpoints 6
    Context in the Discourse 6
    Formal Dissent 8
    Red Team Critique 9
    MAGA Perspective 10
Notable Passages 12
Page-by-Page Summaries 17

Abstracts

TL;DR (One Word)

Subterranean.

Explain It To Me Like I'm Five Years Old

This book talks about fighting underground in wars. It says that it's becoming more and more important for the United States military to know how to fight in underground places. The book tells stories from the past about fighting underground and why it's still important today. It also says that the United States military doesn't focus enough on training and having the right equipment to fight underground. The book suggests that the military needs to change how they think about fighting underground and come up with new ways to find and defeat enemies.

TL;DR (Vanilla)

This book argues that subterranean warfare is becoming increasingly important in modern conflicts and calls for the United States military to develop the necessary forces, training, equipment, and capabilities to effectively combat underground threats. It discusses historical examples and emphasizes the lack of focus on subterranean operations in the US military. The book suggests changes in concepts, capabilities, and technologies to detect and destroy underground threats efficiently. It also draws insights from Israel\u0027s specialized unit for subterranean warfare and proposes integrating various skill sets into task organized teams. Overall, it highlights the importance of being prepared for subterranean environments to overcome challenges and succeed in future conflicts.

SCIENTIFIC STYLE

This thesis discusses the increasing importance of subterranean warfare in modern conflicts and the need for the United States military to develop the necessary forces, training, equipment, and capabilities to effectively combat these threats. The lack of focus, doctrine, training, and equipment in the United States military regarding subterranean operations is emphasized. Historical examples of subterranean warfare are provided, highlighting its relevance in the 21st century. The Israeli Defense Force's specialized unit for subterranean warfare is also mentioned, with insights drawn from their experiences. The integration of various skill sets and capabilities into task organized subterranean warfare teams is suggested. The article concludes by emphasizing the importance of being manned, trained, and equipped to thrive in a subterranean environment in order to overcome the challenges posed by subterranean threats and achieve success in future conflicts.

ACTION ITEMS

Conduct a comprehensive review of current doctrine, training, and equipment related to subterranean warfare in the United States military.

Develop a specialized unit or units within the military dedicated to subterranean warfare, similar to the Israeli Defense Force's model.

Invest in research and development of technologies specifically relevant to winning fights underground.

VIEWPOINTS

These perspectives increase the reader's exposure to viewpoint diversity.

CONTEXT IN THE DISCOURSE

Major Brian T. Spillane's thesis on "Tunnel Rats Version 2.0" delves into the critical subject of subterranean warfare, a domain that has grown in significance in modern warfare. This specialized form of combat is defined by operations conducted beneath the ground surface, involving combat in tunnels, underground facilities, and various subterranean cavities. Its importance is underscored by the fact that many nations possess underground military facilities that are considered vital terrain in contemporary conflicts, and the unique challenges these environments pose to conventional military forces that typically rely on tanks and air support.

Historically, subterranean warfare has been employed as a tactic by less equipped forces to counter stronger adversaries, and its lineage traces back to ancient and medieval times. However, in recent years, there has been a resurgence in the utilization of this form of warfare, with instances such as the Aleppo tunnel bombs and the Israel-Gaza conflict illustrating its current relevance. Moreover, the complex tunnel systems used in recent conflicts, like those in Ukraine, highlight the evolving nature of subterranean threats.

Spillane's thesis is situated within a broader discourse that acknowledges the gap between the current military focus and the need for comprehensive doctrines and training tailored to subterranean combat. This gap is not merely tactical but strategic, with calls for a more refined operational approach to subterranean threats. It involves accurate risk assessments, understanding the impact of tunnels on broader missions, and anticipating the innovative uses of tunnel warfare by various actors.

Furthermore, the increase in positional warfare within dense urban areas and the dissemination of tunneling knowledge among nonstate actors have heightened the relevance of subterranean warfare. The United States military's recognition of these challenges is evidenced by initiatives like the DARPA Subterranean Challenge, which seeks technological solutions to underground warfare through competitive innovation.

In this context, Spillane's thesis contributes to the modern discourse on war by emphasizing the imperative for the U.S. military to evolve and adapt to subterranean challenges. It underlines the strategic necessity for dedicated forces, equipped with the requisite training and technology, to effectively counter subterranean threats. The insights from the Israeli Defense Force's experience in this realm further reinforce the thesis's call for an integrated skill set and the formation of specialized subterranean warfare teams. By highlighting these aspects, the thesis reinforces the need for a paradigm shift in military preparedness for future conflicts where the subterranean domain is likely to play a pivotal role.

SELECTED SOURCES

Collins, John Spencer, Liam. "Underground Nightmares: Preparing for the Future Subterranean Battlefield." Modern War Institute, August 26, 2019. https://mwi.westpoint.edu/underground-nightmares-preparing-future-subterranean-battlefield/.

Richemond-Barak, Daphné, and Stefan Voiculescu-Holvad. "The Rise of Tunnel Warfare as a Tactical, Operational, and Strategic Issue." *Studies in Conflict & Terrorism* 0, no. 0 (2023): 1–20. https://doi.org/10.1080/1057610X.2023.2244191

Soldier of Fortune Magazine. "War in the Tunnels: When Combat Goes Underground," October 10, 2023. https://sofmag.com/war-in-the-tunnels/.

Spencer, John. "The Return of the Tunnel Bomb: A Medieval Tactic on the Modern Battlefield." Modern War Institute, December 30, 2019. https://mwi.westpoint.edu/return-tunnel-bomb-medieval-tactic-modern-battlefield/.

"The Elephant in the Tunnel: Preparing to Fight and Win Underground - Modern War Institute." Accessed November 1, 2023. https://mwi.westpoint.edu/elephant-tunnel-preparing-fight-win-underground/.

www.army.mil. "Interactive Notes from the Underground." Accessed November 1, 2023. https://www.army.mil/article/224511/interactive_notes_from_the_underground.

Formal Dissent

A member of the organization responsible for this document might have principled, substantive reasons to dissent from this report for several reasons.

Resource Allocation: One possible reason for dissent could be the disagreement with the allocation of resources towards subterranean warfare. The dissenting member may argue that the emphasis on subterranean warfare fails to prioritize other important areas of conflict or defense strategies. They might believe that allocating significant resources towards combating subterranean threats could divert resources from addressing more pressing concerns, such as cyber warfare or emerging technologies.

Cost-Effectiveness: Another reason for dissent could be concerns about the cost-effectiveness of investing in subterranean warfare capabilities. The dissenting member might argue that the financial investment required to develop specialized forces, training, and equipment for subterranean operations may outweigh the potential benefits. They may contend that the likelihood of facing subterranean threats in future conflicts is low compared to other types of warfare, making it a less cost-effective pursuit.

Strategic Focus: The dissenting member may question the strategic focus on subterranean warfare, believing that it does not align with the current national security priorities or the organization's overall mission. They may argue that the emphasis on subterranean operations detracts from more critical areas, such as intelligence gathering, counter-terrorism efforts, or diplomatic initiatives. They might advocate for a broader strategic approach that encompasses a range of threats rather than heavily focusing on one specific domain.

Ethical Considerations: Dissent could also arise from ethical concerns regarding subterranean warfare. The member may hold principled objections to the use of certain tactics, techniques, and technologies employed in underground conflicts. They might argue that engaging in subterranean warfare could result in civilian casualties, destruction of historical sites, or violation of international laws and conventions. Their dissent may stem from a belief that pursuing subterranean warfare goes against their moral compass or the values of the organization.

Lack of Evidence: Lastly, a dissenting member might question the evidence presented in the report. They may argue that the historical examples provided do not sufficiently demonstrate the significance or prevalence of subterranean warfare in modern conflicts. They could assert that the lack of concrete data or real-world scenarios undermines the justification for prioritizing resources towards combating subterranean threats. Their dissent may stem from a desire for more robust evidence before making substantial changes to military strategy and capabilities.

In conclusion, a member of the organization responsible for this document might have principled, substantive reasons to dissent from this report due to concerns over resource allocation, cost-effectiveness, strategic focus, ethical considerations, and the lack of evidence supporting the emphasis on subterranean warfare.

RED TEAM CRITIQUE

The main issue with this document is its lack of a critical analysis of the arguments and evidence presented. The author seems to make sweeping statements about the increasing importance of subterranean warfare without providing sufficient evidence or examples to support these claims.

Additionally, the document fails to address potential counterarguments or alternative perspectives. It does not consider the possibility that subterranean warfare may not be as important as other forms of warfare in modern conflicts, or that the resources required to develop forces, training, equipment, and capabilities for subterranean warfare could be better allocated elsewhere.

Furthermore, there is a lack of depth in the historical examples provided. The document mentions historical examples of subterranean warfare but does not provide sufficient analysis or context to understand their relevance in the 21st century. Without a more detailed examination of past conflicts and their lessons learned, it is difficult to evaluate the importance of subterranean warfare today.

The document also fails to acknowledge potential limitations or challenges associated with implementing changes in concepts, capabilities, and technologies for subterranean warfare. It does not address how these changes would be funded or how long it would take to develop and implement them. Without considering these practical aspects, the feasibility of the proposed changes is questionable.

Moreover, the reliance on insights from the Israeli Defense Force's specialized unit for subterranean warfare raises concerns about the applicability of these insights to the United States military. Different countries have different military doctrines, resources, and strategic considerations, so it may not be appropriate to directly apply lessons from one country to another. The document should have provided a more nuanced analysis of the differences and similarities between the Israeli Defense Force and the United States military.

Overall, this document could benefit from a deeper analysis of the arguments and evidence presented, as well as a more comprehensive evaluation of the potential limitations and challenges associated with developing the forces, training, equipment, and capabilities for subterranean warfare.

MAGA Perspective

This book is just another example of liberal fear-mongering and an attempt to undermine our military's focus on more pressing, above-ground threats. Subterranean warfare? Really? Are we supposed to believe that this is the greatest danger our military faces in the 21st century? It is typical of the weak-minded liberals to obsess over these niche topics instead of focusing on the real challenges facing our nation.

The idea of reallocating resources and manpower to combat underground threats is a complete waste. Our military should be focused on defending our borders from illegal immigrants and protecting American jobs from outsourcing. This obsession with subterranean warfare only serves to distract us from the real battles we need to fight.

And why are we looking to the Israeli Defense Force for guidance on this matter? Last time I checked, we were the United States of America, not Israel. We should be relying on our own military strength and expertise, not trying to mimic another country's tactics.

Furthermore, it is concerning that this book suggests changes in concepts, capabilities, and technologies to combat underground threats. This implies that our current strategies and equipment are inadequate, which is simply untrue. Our military has proven time and time again that it is the strongest and most capable in the world. We don't need to waste time and resources on developing new technologies for a problem that doesn't even exist.

Overall, this book is just another example of liberal intellectuals trying to push their own agenda onto our military. We should be wary of anyone who tries to divert our attention away from the real threats to our nation and our values.

NOTABLE PASSAGES

> *"Subterranean operations have long been utilized in warfare and are increasingly prevalent in modern conflicts, as technology proliferation has forced adversaries of the United States underground. However, the United States military currently has no focus, limited doctrine or techniques, tactics, and procedures, scant training, and almost no equipment or technological capabilities to enable its forces to thrive against an enemy in this type of complex operating environment. In future war, the United States needs the forces, training, equipment and capabilities to detect and destroy subterranean threats efficiently and rapidly, and force the enemy back above ground to an environment where the United States has distinct advantages and can dictate the outcome in combat."*

> *"The United States needs to change how it mans, trains, and equips forces in order to set conditions to thrive and defeat adversaries in subterranean environments in future conflicts. Military forces must train and prepare for subterranean threat environments before deploying. The subterranean environment needs to be treated like jungle, urban, and desert environments and prepared for as such. Military forces must be able to leverage standard collections assets and subterranean-specific equipment sets to enable rapid and accurate detection of subterranean threats, and three-dimensional mapping of the underground systems. The United States has created large advantages in information collection and situational awareness above ground, and similar advantages need to be created and employed to defeat underground threats. Task organized subterranean warfare teams must be developed with combat engineers, Explosive Ordnance Demolition (EOD) technicians, demolitions experts, infantry personnel, and MWD teams that can detect, clear, neutralize, and destroy any remaining subterranean threats that are unable to be effectively neutralized by stand-off weapons, or non-human assets."*

> *"Success in war ultimately depends on the ability to adapt and innovate. The period between wars is crucial in shaping the future outcome. It is during this time that strategies are developed, technologies are advanced, and alliances are formed. The decisions made during these interwar periods can have far-reaching consequences, determining whether a nation will emerge victorious or suffer defeat. It is a time of reflection, analysis, and preparation. The lessons learned from past conflicts must be studied and applied, while new ideas and tactics are explored. The interwar period is a test of a nation's resilience and foresight, as it sets the stage for the next chapter in history. It is a time of uncertainty, but also of opportunity. The choices made in this critical juncture will shape the course of future conflicts and determine the fate of nations."*

"ENABLING CONCEPT FOR MODERN SUBTERRANEAN WARFARE"

> *"The complexities and challenges of fighting and winning against an enemy utilizing subterranean systems left an impression on many of the operational planners and tactical executors who wondered how we as a Marine Corps would deal with this challenging environment. My intent in writing this paper was to highlight the problems posed by subterranean warfare, emphasize its relevance in current and future war, and provide solutions to drive changes that will enhance*

	the warfighting capabilities and readiness of the United States military for future wars."
1	*"Imagine fighting an invisible enemy ‚Äì an enemy that can conceal his movements and weapons, conduct attacks and disappear before being effectively targeted, and survive when engaged with modern standoff weapons. While this military scenario sounds like something out of a bad science fiction movie, especially in an age when modern technology provides unparalleled situational awareness on the battlefield, subterranean warfare is a modern military problem that warfighters need to be concerned with."*
2	*"Union troops that had a background in mining dug a 500 foot tunnel under Confederate lines and detonated over 8,000 pounds of black powder under their fortifications, creating an immense crater and killing 276 Confederate troops in the blast."*
3	*"The British utilized 19 separate tunnel systems to successfully detonate over 5 million pounds of explosives under the German positions, killing an estimated 10,000 Germans in the initial blast and creating 7,000 German prisoners of war who were too shell shocked to fight and surrendered."*
4	*"These historical examples highlight how subterranean warfare has a long and very effective history in warfighting. The underground environment has been utilized for offensive as well as defensive purposes, and has provided an unquantifiable psychological impact that compounds the chaos of war. Though subterranean warfare has been utilized in armed conflict throughout history, subterranean warfare has evolved over time from a simple tactic into a critical component of military strategy seeking to achieve victory over a technologically and conventionally superior adversary."*
5	*"The cave and tunnel complexes in some cases were miles long, thousands of feet deep into the mountain slopes, eight feet wide, and provided ventilation, heating, and lighting."*
6	*"Many preconceived notions of subterranean warfare indicate that it is a tool of the weak to fight against an adversary that is much stronger. There is an image of small, dirt caves, barely large enough to fit an average sized man. This is not the case. Many near-peer adversaries that have highly developed military capabilities are utilizing the underground environment to gain and maintain an advantage."*
7	*"Subterranean warfare is not new to the battlefield, but it is increasing in prevalence, sophistication, and importance to military strategy; therefore, it is a combat environment the United States military needs to be prepared to face and be able to thrive in. Subterranean networks are being developed and utilized by potential adversaries across the range of military operations ‚Äì from near-peer threats that have weapons of mass destruction, to non-state actors and transnational terrorist organizations ‚Äì so it is highly likely that the next wars that the United States fights in will involve a subterranean environment."*
8	*"In modern warfare, the United States needs the forces, training, equipment and capabilities to detect and destroy subterranean threats efficiently and rapidly,*

	and force the enemy back above ground to an environment where the United States has distinct advantages and can impose its will on the enemy."
9	*"Because the underground environment is utilized to conceal actions and intent, detection is vital to thriving against this complex military problem. Detection, as early as possible and ideally before the subterranean network is completed, is the optimal approach to defeating subterranean threats and maintaining a position of advantage."*
10	*"The fact that this is a primary approach to detecting and defeating the subterranean problem is concerning given the fact it is not proactive, not adaptive, and does not limit risk to forces. While tactical readiness and training are essential to successful warfighting, these skills do not place friendly forces in a position of advantage relative to adversaries utilizing subterranean networks."*
11	*"During Operation Cast Lead in 2009, Israel's Military Intelligence Directorate conducted significant intelligence preparation of the battlespace and multi-layered intelligence gathering activities in advance of military operations which resulted in Israel identifying and geolocating over 600 tunnel-related targets in Gaza. When combat operations began, Israel struck the targets throughout their campaign with tunnel targets accounting for 17 percent of all targets assigned to Israel's Air Force. Israel's layered approach to intelligence collection proved to be a highly effective model for countering adversaries that utilize subterranean warfare."*
12	*"In subterranean warfare the ability to detect enemy underground networks enables the destruction and defeat of the threat. Currently there are not many proven solutions to aid military personnel in the detection and destruction of underground threats. This problem has persisted for some time."*
13	*"As it becomes more dangerous to operate on the surface, adversaries will increasingly utilize subterranean warfare as their operational approach. The United States needs to prepare to conduct subterranean warfare in future conflicts and must adjust concepts and capabilities to be able to thrive and defeat adversaries in this type of environment."*
14	*"It is evident from historical and modern examples that the subterranean environment is increasing in relevance to future warfare and presents challenges across all warfighting functions. Command and control becomes problematic as communications equipment becomes degraded, or completely ineffective. Maneuver is restricted to confined, narrow underground spaces where darkness, poor visibility, breathing difficulty, uncertainty, and the inability to bring advanced maneuver and fires to bear contribute to a dangerous and psychologically challenging environment for ground forces."*
15	*"The friction and leveling effect created by the subterranean environment make it a strong strategic approach for adversaries seeking to gain a position of advantage over a stronger, more technologically capable force, and achieve victory. For these reasons, subterranean threats pose problems to warfighters across the range of military operations and all the levels of war."*
16	*"Subterranean threats require integrated capabilities and focused training. Israel has spent over a decade combating what some would argue is a strategic threat to their nation posed by subterranean terror tunnels built by Hamas and*

	Hezbollah. To meet this threat head on, Israel created a specialized unit to take the lead on subterranean warfare. Within the Israeli Defense Force (IDF) Combat Engineering Corps, the Yahalom Unit is an engineering force that is organized specifically to address complex engineering problems in modern warfare."
17	*"The goal with subterranean warfare teams is to have a level of proficiency, skill, and equipment that can be brought to bear to enable the clearance and destruction of subterranean threats without causing maneuver elements to lose tempo or initiative, and without placing friendly forces at excessive risk."*
18	*"As the military rewrites doctrine and warfighting publications to account for changes to modern warfare, subterranean operations must be rewritten in order to drive changes to task organization, training, requirements, and capability sets that will bring clarity to subterranean operations and enhance warfighting capability to face these threats in combat."*
19	*"The most critical aspect of defeating subterranean threats is detection. As with most military problems, there is no 'silver bullet' solution to intelligence, or battlefield awareness. In order to successfully fight and defeat adversaries using subterranean systems, equipment sets with subterranean warfare-specific capabilities need to be developed and acquired. These systems need to be able to detect and map subterranean systems to a level of specificity that enable commanders to make decisions, rapidly address, and continue to drive offensive operations without ceding momentum, or the initiative to the enemy. Subterranean detection systems are broken up into two categories: above ground detection and underground detection."*
20	*"Below ground detection methods are capabilities that are still emerging for military purposes, but once realized can enable the United States to gain a significant advantage in fighting in a subterranean environment."*
21	*"After the data is collected and compiled, the technology provides a cross section of what the underground environment looks like."*
22	*"There is no clear-cut technological asset that provides the military with an accurate, survivable, and reliable subterranean threat detection capability. However, there is traction in seeking a better solution."*
23	*"Sending in ground forces plays directly into the operational objectives of weaker adversaries that seek to fight on a level playing field where they can inflict casualties on the stronger force, expand the length of conflict, create events that can be exploited in the information environment, and win by not being defeated."*
24	*"Sending machines into subterranean environments to map, identify, confirm, and destroy underground complexes is preferable to sending ground forces down into a challenging, dangerous, and unknown environment."*
25	*"The United States needs to prepare to conduct subterranean warfare in future conflicts and must adjust concepts and capabilities to be able to thrive and defeat adversaries in this type of environment. The subterranean environment will be present on battlefields of the future. Just like the United States would find it unacceptable to send troops into a jungle environment without being familiar, or equipped to operate in the jungle, it is similarly unacceptable to look at historical*

	examples and recent conflicts and not develop the capabilities to operate and thrive in and around subterranean systems."
26	*"The subterranean threat is not new and is not going to go away. What military forces do in between wars will determine whether or not it is an environment that leads to victory, or defeat."*
27	*"Reimagining the Character of Urban Operations for the US Army: How the Past Can Inform the Present and Future."*
28	*"Securing North Korean Nuclear Sites Would Require Ground Invasion, Pentagon Says."*
29	*"Israeli Official Bets Advances in Anti-Tunnel Technology Will Secure Gaza Border."*[1]
30	*"Digging into Israel: The Sophisticated Tunneling Network of Hamas."*

---

[1] [!—Ed.]

Page-by-Page Summaries

	The page provides information about a report on subterranean warfare and the need for the United States military to develop the necessary forces, training, equipment, and capabilities to effectively combat threats in underground environments. The report highlights the lack of focus, limited doctrine, and insufficient resources currently available for subterranean operations.
	This page is a submission for a Master of Military Studies degree, focusing on fighting and winning future wars in a subterranean environment. The author is Major Brian T. Spillane of the United States Marine Corps.
	The United States needs to develop the forces, training, equipment, and capabilities to efficiently detect and destroy subterranean threats in modern warfare. Adversaries such as Hezbollah, Hamas, Islamic State, North Korea, Iran, and Russia are increasingly using underground environments for military operations. The US military currently lacks focus, doctrine, training, and equipment for subterranean warfare. Changes are needed in how forces are manned, trained, and equipped to thrive and defeat adversaries in subterranean environments. This
	The page discusses how the period between wars can either lead to victory or defeat, emphasizing its importance in determining the outcome.
	This page provides an introduction to subterranean warfare, discussing its historical context, modern relevance, and current methods to combat underground threats. It also presents an enabling concept for modern subterranean warfare, including its purpose, military problem, central idea, manning/task organization, doctrine and training, detection/intelligence, and maneuver. The page concludes with a bibliography.
	Subterranean warfare is an increasing concern for the United States, as weaker adversaries use the underground environment to their advantage. The US lacks focus, doctrine, training, and equipment to effectively combat this type of warfare. Detecting and destroying subterranean threats efficiently is crucial for modern warfare.
1	Subterranean warfare has a long history, with examples dating back to ancient civilizations. It has been used strategically in various conflicts, such as the American Civil War and World War I. The Battle of the Crater and the Battle for Messines Ridge are highlighted as examples.
2	Subterranean warfare has been utilized throughout history, from World War I to Vietnam. Tunnels and underground networks were used for strategic advantage, allowing for surprise attacks, protection, and the ability to move

	undetected. It proved effective in slowing down enemy advances and inflicting casualties.
3	Subterranean warfare has a long history and has evolved into a critical component of military strategy. In the 21st century, advancements in technology have increased the relevance of underground networks in modern warfare.
4	The page discusses the challenges and significance of subterranean warfare in modern conflicts, specifically highlighting the use of caves and tunnels by the Taliban, al Qaeda, Hezbollah, and Hamas. It emphasizes the need for analysis in order to address the military problem effectively.
5	Adversaries such as Hezbollah, Hamas, North Korea, Iran, and Russia have developed extensive subterranean networks for warfare purposes, including tunnels, bunkers, and underground facilities. These networks provide advantages such as avoiding detection, storing weapons, and conducting cross-border attacks. The notion that subterranean warfare is only used by weaker forces is incorrect, as even highly developed military powers utilize these underground environments.
6	Subterranean warfare is increasing in prevalence and importance to military strategy as countries like Russia, North Korea, and Iran utilize underground facilities to conceal their actions, conduct tests, and protect critical assets. The United States military needs to be prepared to face and thrive in this combat environment.
7	The United States needs to develop the necessary forces, training, equipment, and capabilities to effectively combat subterranean threats in modern warfare. Currently, there is a lack of attention, training, doctrine, and techniques in this area. The analysis focuses on current methods to combat these threats in non-classified environments.
8	Detection is crucial in combating subterranean threats, with traditional field craft, military skills, intelligence assets, and technology being employed. Trained personnel, awareness of surroundings, and identifying indicators such as spoil or changes in patterns of life are key in identifying subterranean networks.
9	The primary approach to detecting and defeating subterranean threats is not proactive, adaptive, or effective at limiting risks. Intelligence collection assets, such as unmanned aerial systems and overhead imagery, are useful in detecting subterranean operations. Measurement and signature intelligence (MASINT) can identify patterns of foot traffic that indicate underground systems. Human intelligence (HUMINT) is also valuable.
10	Intelligence collection, including signals intelligence and technology, can effectively detect and counter subterranean threats. Israel's layered approach during Operation Cast Lead successfully identified and targeted tunnel-related targets. Science and technology have been utilized by Israel and the US Department of Homeland Security to detect underground tunnels.
11	Current technology and systems have not been successful in detecting underground tunnels for military purposes. The lack of proven solutions, along

	with a lack of doctrine, training, and procedures, hinders the ability to combat subterranean threats effectively.
12	The United States military must prepare for subterranean warfare as adversaries seek to undermine surface advantages. Solutions are needed for the challenges of operating in underground environments. DARPA is seeking breakthrough technologies for national security.
13	The enabling concept for modern subterranean warfare aims to equip and train the US military to effectively combat enemies using underground environments. Subterranean warfare poses challenges in communication, maneuverability, and intelligence gathering. The objective is to force the enemy back above ground where the US has an advantage.
14	The United States military is currently ill-equipped to handle the challenges posed by subterranean warfare, which can impact logistics, troop protection, and information gathering. Changes in doctrine, training, organization, and equipment are necessary to overcome these challenges and stay ahead of adversaries.
15	Israel's Yahalom Unit, within the IDF Combat Engineering Corps, specializes in addressing subterranean threats posed by Hamas and Hezbollah. The unit, which includes The Samur Unit focused on anti-tunnel warfare, has successfully integrated various capabilities and skill sets to counter these threats while minimizing risk to its forces.
16	Subterranean warfare teams should be flexible, consisting of combat engineers, demolitions experts, EOD technicians, MWD teams, and infantry, with robust robotic capabilities. There is a gap in doctrine and training for subterranean operations that needs to be addressed.
17	Current military publications on subterranean operations lack specific and up-to-date information, failing to address the unique challenges of operating underground. The IDF provides a good starting point for rewriting doctrine and enhancing training. The US military should incorporate subterranean operations into formal courses, unit training, range complexes, and combat readiness evaluations to develop proficiency and identify necessary requirements.
18	Detection is crucial for defeating subterranean threats. Above ground and underground detection methods are needed to detect and map these systems, allowing commanders to make informed decisions and maintain offensive operations. Boots on the ground should be a last resort method. A layered approach to intelligence collection is recommended.
19	Ground-penetrating radar and seismic imaging are emerging technologies that can help detect and locate underground systems. Aviation platforms can also provide valuable insights by observing surface cues. However, these methods have limitations and require further advancements to be effective in combat environments.
20	Private sector and government agencies are developing and improving technologies to map and identify underground resources and threats. These include seismic imaging, unmanned aerial systems with magnetic field sensors, vehicle-mounted seismic imagers, and LIDAR-based systems.

21	The military lacks a reliable subterranean threat detection capability, but Israel and the United States are investing in research and development for anti-tunneling programs. They are developing sensor-equipped walls and testing unmanned vehicles with radar technology to detect underground networks.
22	The page discusses the limitations of current subterranean detection and mapping technology for military use, and the need for breakthrough technologies in this field. It also highlights the challenges of neutralizing underground threats and the ineffectiveness of current methods.
23	Utilizing Robotics and Autonomous Systems (RAS) and Manned, Unmanned Teaming (MUM-T) is the solution to combat subterranean threats, as it avoids sending ground forces into dangerous and unknown environments. The United States military should incorporate these technologies to effectively operate in underground systems.
24	The United States must prepare for subterranean warfare in future conflicts, developing the necessary training, equipment, and capabilities to thrive in this environment. This includes leveraging standard collections assets, subterranean-specific equipment sets, and task organized subterranean warfare teams to detect, clear, neutralize, and destroy subterranean threats. These specialized forces will allow primary maneuver forces to focus on aggression and tempo against the enemy.
25	The United States needs to adapt its military tactics and technologies to effectively combat subterranean threats, relying on remote systems and unmanned vehicles rather than committing ground forces. This will determine success or failure in this challenging environment.
26	This page is a bibliography of various sources related to underground warfare, including reports on Israel's wars in Gaza, the DARPA Subterranean Challenge, and articles on detecting and blocking underground threats.
27	The page contains a list of sources related to military operations in mountain and urban terrains, including articles, books, and government documents.
28	The page contains various sources related to underground and tunnel warfare, including reports on nuclear earth penetrator weapons, Israel's anti-tunnel technology, border patrol techniques for finding smuggler tunnels, and military handbooks on subterranean operations.
29	This page contains references to two articles: one about Hamas' tunneling network in Israel and another about being lost at Tora Bora.

REPORT DOCUMENTATION PAGE

Form Approved
OMB No. 0704-0188

Public reporting burden for this collection of information is estimated to average 1 hour per response, including the time for reviewing instructions, searching existing data sources, gathering and maintaining the data needed, and completing and reviewing this collection of information. Send comments regarding this burden estimate or any other aspect of this collection of information, including suggestions for reducing this burden to Department of Defense, Washington Headquarters Services, Directorate for Information Operations and Reports (0704-0188), 1215 Jefferson Davis Highway, Suite 1204, Arlington, VA 22202-4302. Respondents should be aware that notwithstanding any other provision of law, no person shall be subject to any penalty for failing to comply with a collection of information if it does not display a currently valid OMB control number. **PLEASE DO NOT RETURN YOUR FORM TO THE ABOVE ADDRESS.**

1. REPORT DATE *(DD-MM-YYYY)*	2. REPORT TYPE	3. DATES COVERED *(From - To)*

4. TITLE AND SUBTITLE	5a. CONTRACT NUMBER
	5b. GRANT NUMBER
	5c. PROGRAM ELEMENT NUMBER

6. AUTHOR(S)	5d. PROJECT NUMBER
	5e. TASK NUMBER
	5f. WORK UNIT NUMBER

7. PERFORMING ORGANIZATION NAME(S) AND ADDRESS(ES)	8. PERFORMING ORGANIZATION REPORT NUMBER

9. SPONSORING / MONITORING AGENCY NAME(S) AND ADDRESS(ES)	10. SPONSOR/MONITOR'S ACRONYM(S)
	11. SPONSOR/MONITOR'S REPORT NUMBER(S)

12. DISTRIBUTION / AVAILABILITY STATEMENT

13. SUPPLEMENTARY NOTES

14. ABSTRACT

15. SUBJECT TERMS

16. SECURITY CLASSIFICATION OF:			17. LIMITATION OF ABSTRACT	18. NUMBER OF PAGES	19a. NAME OF RESPONSIBLE PERSON
a. REPORT	b. ABSTRACT	c. THIS PAGE			
					19b. TELEPHONE NUMBER *(include area code)*

Standard Form 298 (Rev. 8-98)
Prescribed by ANSI Std. Z39.18

United States Marine Corps
Command and Staff College
Marine Corps University
2076 South Street
Marine Corps Combat Development Command
Quantico, Virginia 22134-5068

MASTER OF MILITARY STUDIES

TUNNEL RATS VERSION 2.0: FIGHTING AND WINNING FUTURE WAR IN A SUBTERRANEAN ENVIRONMENT

SUBMITTED IN PARTIAL FULFILLMENT OF THE REQUIREMENTS FOR THE
DEGREE OF MASTER OF MILITARY STUDIES

MAJOR BRIAN T. SPILLANE, UNITED STATES MARINE CORPS

AY 2017-18

Mentor and Oral Defense Committee Member: Dr. Benjamin Jensen, Ph.D
Approved: _____
Date: _____

Oral Defense Committee Member: Dr. Benjamin Jensen, Ph.D
Approved: _____
Date: 2/30/18

Oral Defense Committee Member: Dr. Paul Gelpi, Ph.D
Approved: _____
Date: 30 April 18

United States Marine Corps
Command and Staff College
Marine Corps University
2076 South Street
Marine Corps Combat Development Command
Quantico, Virginia 22134-5068

MASTER OF MILITARY STUDIES

TUNNEL RATS VERSION 2.0: FIGHTING AND WINNING FUTURE WAR IN A SUBTERRANEAN ENVIRONMENT

SUBMITTED IN PARTIAL FULFILLMENT OF THE REQUIREMENTS FOR THE DEGREE OF MASTER OF MILITARY STUDIES

MAJOR BRIAN T. SPILLANE, UNITED STATES MARINE CORPS

AY 2017-18

Mentor and Oral Defense Committee Member: Dr. Benjamin Jensen, Ph.D
Approved: _____
Date: _____

Oral Defense Committee Member: Dr. Benjamin Jensen, Ph.D
Approved: _____
Date: _____

Oral Defense Committee Member: Dr. Paul Gelpi, Ph.D
Approved: _____
Date: _____

EXECUTIVE SUMMARY

Title: Tunnel Rats Version 2.0: Fighting and Winning Future War in a Subterranean Environment.

Author: Major Brian T. Spillane, United States Marine Corps.

Thesis: In modern warfare, the United States needs the forces, training, equipment and capabilities to detect and destroy subterranean threats efficiently and rapidly, and force the enemy back above ground to an environment where the United States has distinct advantages and can dictate the outcome in combat.

Discussion: Subterranean operations have long been utilized in warfare and are increasingly prevalent in modern conflicts, as technology proliferation has forced adversaries of the United States underground. In the 21st century, adversaries around the world like Hezbollah, Hamas, Islamic State, North Korea, Iran, and Russia are using the subterranean environment for military operations. The United States military currently has no focus, limited doctrine or techniques, tactics, and procedures, scant training, and almost no equipment or technological capabilities to enable United States forces to thrive against the enemy in this type of complex operating environment. In this critical period between wars, the United States must prepare to conduct subterranean warfare in future conflicts and must adjust concepts and capabilities to be able to thrive and defeat adversaries in this type of environment.

Conclusion: The United States needs to change how it mans, trains, and equips forces in order to set conditions to thrive and defeat adversaries in subterranean environments in future conflicts. Military forces must train and prepare for subterranean threat environments before deploying. The subterranean environment needs to be treated like jungle, urban, and desert environments and prepared for as such. Military forces must be able to leverage standard collections assets and subterranean-specific equipment sets to enable rapid and accurate detection of subterranean threats, and three-dimensional mapping of the underground systems. The United States has created large advantages in information collection and situational awareness above ground, and similar advantages need to be created and employed to defeat underground threats. Task organized subterranean warfare teams must be developed with combat engineers, Explosive Ordnance Demolition (EOD) technicians, demolitions experts, infantry personnel, and MWD teams that can detect, clear, neutralize, and destroy any remaining subterranean threats that are unable to be effectively neutralized by stand-off weapons, or non-human assets. The United States must incorporate low-risk technology, Robotics and Autonomous Systems (RAS), and Manned Unmanned Teaming (MUM-T) that is developed specifically for defeating subterranean threats. Only as a last resort should ground forces be committed underground to fight and defeat the enemy in subterranean systems. RAS and MUM-T should force traditional tunnel rats out of a job. The subterranean threat is not new and is not going to go away. What military forces do in between wars will determine whether or not it is an environment that leads to victory, or defeat.

DISCLAIMER
PREFACE i
I. INRODUCTION 1
II. HISTORICAL CONTEXT 2
III. MODERN RELEVANCE 4
IV. CURRENT METHODS TO COMBAT SUBTERRANEAN THREATS 8
V. ENABLING CONCEPT FOR MODERN SUBTERRANEAN WARFARE 14
    A. Purpose 14
    B. Military Problem 14
    C. Central Idea 15
    D. Manning/Task Organization 16
    E. Doctrine and Training 17
    F. Detection /Intelligence 19
    G. Maneuver 23
VI. CONCLUSION 25
BIBLIOGRAPHY 27

DISCLAIMER

THE OPINIONS AND CONCLUSIONS EXPRESSED HEREIN ARE THOSE OF THE INDIVIDUAL STUDENT AUTHOR AND DO NOT NECESSARILY REPRESENT THE VIEWS OF EITHER THE MARINE CORPS COMMAND AND STAFF COLLEGE OR ANY OTHER GOVERNMENT AGENCY. REFERENCES TO THIS STUDY SHOULD INCLUDE THE FOREGOING STATEMENT.

QUOTATION FROM, ABSTRACTION FROM, OR REPRODUCTION OF ALL OR ANY PART OF THIS DOCUMENT IS PERMITTED PROVIDED PROPER ACKNOWLEDGEMENT IS MADE.

PREFACE

Elements of the Marine Corps operating forces conducted a warfighting exercise in early 2017 and one of the scenarios involved executing offensive operations against a near-peer enemy fighting from subterranean systems. The complex exercise integrated live training with an overarching computer-based simulated scenario. The subterranean environment was part of the simulated world, and while emphasized during threat briefs and the intelligence preparation of the battlespace, was largely overlooked during planning and execution. It was not surprising that the underground threats were overlooked given the fact that there was no experience base, limited doctrine and warfighting references to refer to, and no equipment sets or resources to enhance capabilities in the subterranean environment. The complexities and challenges of fighting and winning against an enemy utilizing subterranean systems left an impression on many of the operational planners and tactical executors who wondered how we as a Marine Corps would deal with this challenging environment. My intent in writing this paper was to highlight the problems posed by subterranean warfare, emphasize its relevance in current and future war, and provide solutions to drive changes that will enhance the warfighting capabilities and readiness of the United States military for future wars.

Numerous individuals helped me throughout the research and writing process for this paper. I would primarily like to thank Dr. Ben Jensen for his guidance and mentorship. His leadership, knowledge, creativity, and enthusiasm encouraged me throughout the generation, research, and writing process, and he greatly influenced the final product. I would also like to thank Major James Geiger and Captain William

Springer from the Ellis Group at the Marine Corps Warfighting Laboratory for their expertise, technical assistance, and support.

While the aforementioned individuals provided invaluable advice during the writing of this paper, the views, opinions, findings, and conclusions expressed in this paper are strictly my own. They are not responsible for any errors or omissions in this paper.

I. INTRODUCTION

Imagine fighting an invisible enemy – an enemy that can conceal his movements and weapons, conduct attacks and disappear before being effectively targeted, and survive when engaged with modern standoff weapons. While this military scenario sounds like something out of a bad science fiction movie, especially in an age when modern technology provides unparalleled situational awareness on the battlefield, subterranean warfare is a modern military problem that warfighters need to be concerned with. In fact, in an age of technology proliferation and advanced precision strike networks, utilizing subterranean warfare is becoming more prevalent and attractive to adversaries of the United States. For conventionally weaker adversaries, subterranean warfare has become a critical component to military strategy. This strategy utilizes the subterranean environment to force technologically advanced and conventionally stronger powers to move away from a stand-off doctrine, increases the length of conflict, forces the commitment of ground forces, increases damages which can be exploited in the information environment, and enables the ability to "win" simply by not losing. Though the United States has faced, and will continue to face adversaries that utilize the underground environment to fight, there is little focus, almost no doctrine or techniques, tactics, and procedures, limited training, and almost no equipment capability sets to enable United States forces to thrive against the enemy in this type of complex operating environment. In modern warfare, the United States needs the forces, training, equipment and capabilities to detect and destroy subterranean threats efficiently and rapidly, and force the enemy back above ground to an environment where the United States has distinct advantages and can dictate the outcome of combat operations.

II. HISTORICAL CONTEXT

Subterranean operations are not new to warfare. Historical examples of subterranean warfare go back centuries to 500 BC as ancient civilizations went underground – even building underground cities – in order to defend themselves and protect their interests against external threats.[1] Recent historical examples are of more interest to modern-day warfighters. From the 19th century to the Cold War, subterranean warfare has been utilized offensively, defensively, and as a strategy to affect the outcome in war. While the means and methods of warfare constantly evolve over time, subterranean warfare has been an effective method of operating over time and across multiple different conflicts. To illustrate this, examples from the American Civil War, World War I, World War II, and Vietnam are highlighted.

The Battle of the Crater in Petersburg, Virginia during the American Civil War in 1864 provides one of the earliest examples of subterranean warfare in American history. Union troops that had a background in mining dug a 500 foot tunnel under Confederate lines and detonated over 8,000 pounds of black powder under their fortifications, creating an immense crater and killing 276 Confederate troops in the blast.[2]

The Battle for Messines Ridge in Belgium during World War I in 1917 is an example of a mining-countermining battle. With the German forces occupying key

---

[1] Donald M. Helig, "Subterranean Warfare: A Counter to U.S. Airpower," (masters thesis, Air Command and Staff College, 2000), 3.
[2] Stew Magnuson, "Holding the Low Ground: Daunting Challenges Face Those Waging Subterranean Warfare," *National Defense*, 91, 639 (2007): 20, https://search-proquest-com.lomc.idm.oclc.org/docview/213305222?OpenUrlRefId=info:xri/sid:wcdiscovery&accountid=14746

terrain, the British dug over five miles of tunnels, some over 100 feet underground, while the Germans conducted countermining efforts aimed at detecting and destroying the British subterranean efforts. The British utilized 19 separate tunnel systems to successfully detonate over 5 million pounds of explosives under the German positions, killing an estimated 10,000 Germans in the initial blast and creating 7,000 German prisoners of war who were too shell shocked to fight and surrendered.[3]

During Word War II the Japanese made extensive use of subterranean networks in an attempt to fight off the Allied advance in the Pacific. Most notably were the Japanese integrated defenses at Peleliu, Iwo Jima, and Okinawa that featured extensive and sophisticated tunnel systems used for protection, maneuver, and advantageous fighting positions.[4] While the Japanese eventually lost these battles, their use of subterranean networks slowed the Allied advance, made superior American fire support assets ineffective, and inflicted massive casualties on the American forces.[5]

During the Vietnam War, America saw firsthand the multi-dimensional threat posed by subterranean warfare. The Viet Cong utilized hundreds of kilometers of underground tunnels that spread from southern Vietnam to Cambodia. The complex tunnel system was started years before the Vietnam War, and enabled the Viet Cong to move fighters and supplies undetected, conduct surprise attacks, and egress without a trace. Additionally, the tunnel system gave the Viet Cong protection from superior

---

[3] Wayne Dillon, "Subterranean Warfare Considerations," (draft report, US Marine Corps Tactics and Operations Group, 2015), 4.

[4] Stew Magnuson, "Holding the Low Ground: Daunting Challenges Face Those Waging Subterranean Warfare," *National Defense*, 91, 639 (2007): 20-21, https://search-proquest-com.lomc.idm.oclc.org/docview/213305222?OpenUrlRefId=info:xri/sid:wcdiscovery&accountid=14746

[5] Donald M. Helig, "Subterranean Warfare: A Counter to U.S. Airpower," (masters thesis, Air Command and Staff College, 2000), 4, 11-15.

American firepower, and provided a constant psychological weapon that created fear and uncertainty, and negatively impacted morale.[6]

These historical examples highlight how subterranean warfare has a long and very effective history in warfighting. The underground environment has been utilized for offensive as well as defensive purposes, and has provided an unquantifiable psychological impact that compounds the chaos of war. Though subterranean warfare has been utilized in armed conflict throughout history, subterranean warfare has evolved over time from a simple tactic into a critical component of military strategy seeking to achieve victory over a technologically and conventionally superior adversary. As modern warfare has changed, subterranean warfare has become even more relevant to modern-day warfighters seeking to prepare for future war.

III. MODERN RELEVANCE

There has been an increase in subterranean warfare and an expansion in the sophistication of underground networks in the 21st century. This is the result of warfighting organizations that are adapting to changes in modern warfare. Armed and unarmed unmanned aerial systems, precision munitions, sensor technology, signals intelligence, electronic and information warfare, and enhanced command and control equipment have increased and are readily accessible to more groups and actors, not just large states. This proliferation of capabilities has enabled detection and lethal targeting almost anywhere on the battlefield, day or night, with precision and speed. As a result, Nations, actors, and/or groups that anticipate conducting military operations have developed and increased the use of subterranean facilities in order to gain an advantage

---

[6] *Ibid.*, 6-9.

on the battlefield. In order to identify the military problem and prove relevance, analysis of modern subterranean warfare is required.

After the September 11th attacks in 2001, the United States went to war in Afghanistan fighting the Taliban, al Qaeda and its leader Osama Bin Laden. The Taliban and Al Qaeda forces that survived the fall of Kandahar retreated to the Tora Bora Mountains where they had well established cave and tunnel complexes that provided significant protection from American airpower. The cave and tunnel complexes in some cases were miles long, thousands of feet deep into the mountain slopes, eight feet wide, and provided ventilation, heating, and lighting.[7] While the United States had a tremendous conventional capability advantage relative to the Taliban and al Qaeda fighters, the caves presented a significant challenge to defeat, and forced the United States to make a choice between committing ground forces and risking casualties, or sticking with stand-off weapons and potentially allowing al Qaeda leaders to survive and get away.

Some of the best After Action Reviews and reports of modern day subterranean warfare come from recent conflicts between Israel and Hezbollah during the Second Lebanon War in 2006, and between Israel and Hamas in 2009, 2012, and 2014. Prior to and during these conflicts Hezbollah and Hamas adapted to the capability disadvantage that they had relative to the Israelis. They made subterranean warfare a central part of their strategy to defeat Israel. Already having historical smuggling tunnels in place, Hezbollah and Hamas took what was there, built new tunnels, and enhanced their

---

[7] Benjamin S. Lambeth, *Air Power Against Terror: America's Conduct of Operation Enduring Freedom*, (RAND Corporation, 2001), 145-152. ProQuest Ebook Central, http://ebookcentral.proquest.com/lib/usmcu-ebooks/detail.action?docID=618737.

subterranean capability. Hezbollah and Hamas had hundreds of kilometers of subterranean tunnels that were in some cases 90 feet deep and reinforced with concrete. Many of the tunnels had electricity, bathrooms, communications links, and additional protection for leadership and strategic command and control sites. Both Hezbollah and Hamas effectively utilized their subterranean network to avoid detection, provide maneuverability for attacks, to store weapons and supplies, and to impact their adversary psychologically by conducting cross border attacks targeting Israeli citizens.

Many preconceived notions of subterranean warfare indicate that it is a tool of the weak to fight against an adversary that is much stronger. There is an image of small, dirt caves, barely large enough to fit an average sized man. This is not the case. Many near-peer adversaries that have highly developed military capabilities are utilizing the underground environment to gain and maintain an advantage. North Korea realized the value to operating underground decades ago, and began a fortification program in 1962 to place most of its critical military infrastructure deep underground.[8] These underground facilities are a critical component to North Korea's military strategy against the United States to the present day. South Korea has even exposed numerous underground tunnels that were cut under the De-Militarized Zone (DMZ) to provide avenues of approach for North Korea to attack into South Korea. North Korea and Iran have sophisticated underground facilities and have enhanced their underground networks after studying the playbook used by the United States military. Russia has a large underground system created during the war that features tunnels, subway systems, bunkers, and military

---

[8] US Army Asymmetric Warfare Group, *Subterranean Operations (SbTo) Handbook V3*, Asymmetric Warfare Group, (Fort Meade, Virginia, August 12, 2015), 6.

facilities.⁹ Iran, like Russia and North Korea, has placed critical military assets and infrastructure in deep underground facilities. North Korea and Iran utilize underground facilities to conceal their actions, conduct military tests, protect critical weapon systems and military assets, and provide hardened positions to conduct attacks and defend from in the case of war.¹⁰

The trend of increased utilization of subterranean warfare continues with the most current, and most followed battles of the last year. In 2017, the Islamic State of Iraq and Syria (ISIS) utilized underground networks in both the Battle of Mosul in Iraq, and in the Battle of Marawi in the Philippines.¹¹ Like other modern examples, the tunnels were pre-planned, well developed, and part of ISIS' overall warfighting strategy.

Subterranean warfare is not new to the battlefield, but it is increasing in prevalence, sophistication, and importance to military strategy; therefore, it is a combat environment the United States military needs to be prepared to face and be able to thrive in. Subterranean networks are being developed and utilized by potential adversaries across the range of military operations – from near-peer threats that have weapons of mass destruction, to non-state actors and transnational terrorist organizations – so it is highly likely that the next wars that the United States fights in will involve a subterranean environment. However, in the United States military subterranean warfare receives little

---

⁹ US Army Training Circular 3-21.50, *Small Unit Training In Subterranean Environments, US Army*, (Department of the Army: Washington, DC, November 2017), 1-2.

¹⁰ Dan Lemothe and Carol Morello, "Securing North Korean Nuclear Sites Would Require Ground Invasion, Pentagon Says," *Washington Post*, last modified November 4, 2017, https://www.washingtonpost.com/world/national-security/securing-north-korean-nuclear-sites-would-require-a-ground-invasion-pentagon-says/2017/11/04/32d5f6fa-c0cf-11e7-97d9-bdab5a0ab381_story.html?utm_term=.f07ce4940691

¹¹ Benjamin Hall, "Exclusive: Inside ISIS' Extensive Tunnel System," (video), *Fox News*, last modified October 23, 2016, http://www.foxnews.com/world/2016/10/23/exclusive-inside-isis-extensive-tunnel-system.html

to no attention or training, and there is currently a deficiency in doctrine and techniques, tactics, and procedures associated with subterranean operations. In modern warfare, the United States needs the forces, training, equipment and capabilities to detect and destroy subterranean threats efficiently and rapidly, and force the enemy back above ground to an environment where the United States has distinct advantages and can impose its will on the enemy. Subterranean operations pose a unique and challenging problem to warfighters.

IV. CURRENT METHODS TO COMBAT SUBTERRANEAN THREATS

Subterranean warfare is a military problem that the United States will face in future armed conflicts. This is problematic because the United States has little focus, almost no doctrine or techniques, tactics, and procedures, and limited training to enable United States military forces to thrive against the enemy in this type of complex operating environment. After examining the historical context and modern day examples, it is important to identify methods that have been and are being utilized to combat subterranean threats.

In order to focus the analysis and make it relevant to the widest audience of military professionals possible, the scope of assessing current methods to combat subterranean threats has been narrowed to focus on threat environments that the United States is likely to be operating within in the near future, and topics that can be discussed at the unclassified level. Therefore, there is no discussion on classified capabilities or plans to defeat deep hardened underground facilities that near-peer competitors of the United States may have.

Defeating subterranean threats involves two parts – detecting the threat, and destruction of the threat.[12] Because the underground environment is utilized to conceal actions and intent, detection is vital to thriving against this complex military problem.[13] Detection, as early as possible and ideally before the subterranean network is completed, is the optimal approach to defeating subterranean threats and maintaining a position of advantage. Detection methods currently being employed by warfighters combating subterranean threats are traditional field craft and basic military skills, intelligence collections assets, and technological capabilities.

Traditional field craft and basic military skills refer to general purpose forces on the ground using the human and physical terrain to identify indications and warnings that adversaries are creating, or operating from the underground environment. The few military publications in the United States that discuss subterranean operations all highlight these tactical skills more than any other capability to counter the threat.[14] Having personnel that are trained observers, know their battle space, and maintain awareness of what is going on around them is one method of identifying subterranean networks. Indicators that can be identified by field craft and basic military skills are identifying spoil, out of place smells, indicators of construction in areas where nothing is being built on the surface, pattern of life changes, anomalies to normal activities in urban

---

[12] Yiftah S. Shapir and Gal Perel, "Subterranean Warfare: A New-Old Challenge," The Institute for National Security Studies, Tel Aviv Univeristy (UNK), 53. http://www.inss.org.il/he/wp-content/uploads/sites/2/systemfiles/SystemFiles/Subterranean%20Warfare_%20A%20New-Old%20Challenge.pdf

[13] *Ibid.*

[14] Publications referenced included: US Army Asymmetric Group Subterranean Operations Handbook (2015), US Army Tactics Techniques Procedures 3-21.50 Infantry Small-Unit Mountain Operations (2011), Marine Corps Warfighting Publication 3-17.4 Engineer Reconnaissance (2016), and Marine Corps Reference Publication 12-10B.1 Military Operations on Urbanized Terrain (2016).

areas, and points of origin for incoming direct and indirect fire.[15] All of these individual tactical actions while important, are implied tasks to military members and not specific to combating subterranean threats. The fact that this is a primary approach to detecting and defeating the subterranean problem is concerning given the fact it is not proactive, not adaptive, and does not limit risk to forces. While tactical readiness and training are essential to successful warfighting, these skills do not place friendly forces in a position of advantage relative to adversaries utilizing subterranean networks.

Intelligence collections assets have been an important source of detecting and defeating subterranean networks. Unmanned aerial systems, aviation platforms, and other overhead assets provide real-time video and imagery intelligence (IMINT) that has proven extremely useful to detecting subterranean operations, especially as they are being constructed. Having a much larger and clearer aperture than individuals on the ground, imagery intelligence can identify indicators, monitor adversary actions, and can also locate heat signatures that help in detecting subterranean networks. Measurement and signature intelligence (MASINT) is used to identify, track, and describe signatures of fixed, or dynamic targets.[16] One of the less discussed intelligence collection capabilities, MASINT is beneficial to identify patterns from adversaries that are seeking to avoid compromise from known overhead imagery capabilities; thus, it is very useful for combating subterranean operations. MASINT can monitor and detect patterns of foot traffic in and around buildings or terrain features that can provide cues to underground systems being built and utilized by enemy forces. Human intelligence (HUMINT) and

---

[15] *Ibid.*
[16] John D. Macartney, "John, How Should We Explain MASINT?" in *Intelligence and the National Security Strategist: Enduring Issues and Challenges*, ed. Roger Z. George and Robert D. Kline (Lanham, MD: Rowman and Littlefield Publishers, 2006), 169.

signals intelligence (SIGINT) are also beneficial methods of identifying subterranean operations by getting information directly from sources with direct knowledge of adversary actions about the locations and activities of enemy forces utilizing subterranean systems. The major drawback to intelligence collection is that assets are limited and competing with other priorities. However, when utilized properly, the impact of layered collections can have a significant impact on detecting and defeating subterranean threats. During Operation Cast Lead in 2009, Israel's Military Intelligence Directorate conducted significant intelligence preparation of the battlespace and multi-layered intelligence gathering activities in advance of military operations which resulted in Israel identifying and geolocating over 600 tunnel-related targets in Gaza.[17] When combat operations began, Israel struck the targets throughout their campaign with tunnel targets accounting for 17 percent of all targets assigned to Israel's Air Force.[18] Israel's layered approach to intelligence collection proved to be a highly effective model for countering adversaries that utilize subterranean warfare.

Science and technology can be a significant force multiplier against challenging military problems. Israel, fighting Hamas and Hezbollah in the Middle East, and the United States Department of Homeland Security (DHS), combating drug and smuggling tunnels along the southwest border of the United States, have tried and tested a range of assets to detect underground tunnels such as geophones, radar, and various other collections assets.[19] Many nations, including the United States and Israel, have invested

---

[17] Benjamin S. Lambeth, *Air Operations in Israel's War Against Hezbollah: Learning from Lebanon and Getting It Right in Gaza*, (RAND Corporation, 2011). ProQuest Ebook Central, http://ebookcentral.proquest.com/lib/usmcu-ebooks/detail.action?docID=744531.
[18] *Ibid.*
[19] *Ibid.*

millions of dollars into anti-tunnel technology development and continue to seek technological solutions to the subterranean threat.[20] While many technologies exist and have been used successfully for civilian purposes, the results in detecting subterranean features for military purposes have been insignificant. In Israel at least four different, expensive systems were developed, but failed to detect underground tunnels.[21] Similarly, a few years ago, a DHS representative stated that all of the cross border tunnels that had been detected by the United States were found through good law enforcement, or chance. He continued, "None by technology."[22]

From the analysis above, two points stand out: First, in subterranean warfare the ability to detect enemy underground networks enables the destruction and defeat of the threat. Second, currently there are not many proven solutions to aid military personnel in the detection and destruction of underground threats. This second point, coupled with a lack of doctrine, training, and techniques, tactics, and procedures, does not establish conditions for success against enemy's that utilize subterranean warfare. This problem has persisted for some time. In 2007, General John Abazaid, former Commander of United States Central Command (CENTCOM), was not happy with the military's ability to detect underground passageways and stated, "On a scale of one to 10, the technology is

---

[20] Raphael S. Cohen, David E. Johnson, David E. Thaler, Brenna Allen, Elizabeth M. Bartels, James Cahill, Shira Efron, *From Cast Lead to Protective Edge: Lessons from Israel's Wars in Gaza.* RAND Corporation (Santa Monica, CA: RAND Corporation, 2017), 161.
[21] Yiftah S. Shapir and Gal Perel, "Subterranean Warfare: A New-Old Challenge," The Institute for National Security Studies, Tel Aviv Univeristy (UNK), 56. http://www.inss.org.il/he/wp-content/uploads/sites/2/systemfiles/SystemFiles/Subterranean%20Warfare_%20A%20New-Old%20Challenge.pdf
[22] US Department of Homeland Security. *Tunnel Vision* (Washington, DC, 2009) Accessed on February 2, 2018, https://www.dhs.gov/science-and-technology/tunnel-vision

a four. We need more ability to see underground."[23] Not much has changed in over a decade. In 2017, the Defense Advanced Research Projects Agency (DARPA) put out a Request for Information, which stated, "In many ways, subterranean environments have remained an untapped domain in terms of developing breakthrough technologies for national security. We're looking for audacious ideas on how to overcome the multi-faceted challenges these locations present…and provide previously unimaginable capabilities for warfighters and emergency responders."[24]

As far as the United States military and preparing for future war, the subterranean environment has simply been ignored. This is concerning. The enemies of the United States have been watching, learning, and adapting. They are well aware that the United States has advantages in aviation technology, precision fires, and intelligence, reconnaissance, and surveillance capabilities. These adversaries seek opportunities to undercut these advantages, and will do everything in their power to wage future war on a more level playing field, or one tilted in their favor. As it becomes more dangerous to operate on the surface, adversaries will increasingly utilize subterranean warfare as their operational approach. The United States needs to prepare to conduct subterranean warfare in future conflicts and must adjust concepts and capabilities to be able to thrive and defeat adversaries in this type of environment. The good news is that there are solutions to these military problems. To thrive in an environment where subterranean operations are employed, the United States needs the forces, training, equipment and

---

[23] Stew Magnuson, "Holding the Low Ground: Daunting Challenges Face Those Waging Subterranean Warfare," *National Defense*, 91, 639 (2007): 20, https://search-proquest-com.lomc.idm.oclc.org/docview/213305222?OpenUrlRefId=info:xri/sid:wcdiscovery&accountid=14746

[24] Defense Advanced Research Project Agency, "DARPA Digging for Ideas to Revolutionize Subterranean Mapping and Navigation," *Defense Advanced Research Project Agency*, accessed November 13, 2017, https://www.darpa.mil/news-events/2017-11-21

capabilities to detect and destroy subterranean threats efficiently and rapidly, and force the enemy back above ground to an environment where the United States has distinct advantages and can impose its will on the enemy. There are numerous ways to accomplish these objectives.

V. ENABLING CONCEPT FOR MODERN SUBTERRANEAN WARFARE

A. Purpose

The purpose of the enabling concept for modern subterranean warfare is to provide a formula to man, train, equip, and operate that gives the United States military the capabilities to thrive against enemy's using subterranean environments and force the enemy back above ground where the United States has distinct advantages and can impose its will on the enemy.

B. Military Problem

It is evident from historical and modern examples that the subterranean environment is increasing in relevance to future warfare and presents challenges across all warfighting functions. Command and control becomes problematic as communications equipment becomes degraded, or completely ineffective. Maneuver is restricted to confined, narrow underground spaces where darkness, poor visibility, breathing difficulty, uncertainty, and the inability to bring advanced maneuver and fires to bear contribute to a dangerous and psychologically challenging environment for ground forces. The employment of many fire support assets becomes useless, as many capabilities and munitions are ineffective against subterranean systems, even if accurately targeted and hit. The enemy becomes harder to identify, detect, and observe, challenging intelligence professionals to understand the threats and provide the commander with information to aid in timely decision-making. Logistics is more time and resource intensive and creates unique challenges to supply lines. Protection for troops is impacted as risk to force increases in subterranean systems that favor the "defender" against a maneuver force that is unfamiliar with the underground environment. In the information environment, the enemy can use the subterranean environment to deceive friendly forces and influence target populations. Some subterranean systems are so deep that robotics and autonomous systems fail to function, or cannot communicate with operators on the surface, limiting technological solutions to the problem. The friction and leveling effect created by the subterranean environment make it a strong strategic approach for adversaries seeking to gain a position of advantage over a stronger, more technologically capable force, and achieve victory. For these reasons, subterranean threats pose problems to warfighters across the range of military operations and all the levels of war. Subterranean threats are only increasing and solutions need to be developed and implemented now to stay ahead of adversaries.

C. Central Idea

To effectively deal with these threats, the United States must be manned, trained, and equipped to thrive in a subterranean environment. However, the doctrine, training, and equipment in the United States military for subterranean warfare is insufficient to meet the challenges posed by facing an enemy in this environment. If changes are not implemented, the United States will be at a disadvantage in future conflicts. This may lead to unfavorable outcomes in terms of risk to force and mission. To overcome these challenges and thrive in a subterranean combat environment, changes must be made to doctrine, training, organization, and equipment sets in order to enable rapid detection and defeat of underground systems while maintaining tempo and the initiative. Having the capabilities to detect and destroy subterranean threats rapidly would serve to enhance friendly force maneuver, and force the enemy back above ground to an environment where the United States has distinct advantages and can more easily dictate the outcome of combat operations.

D. Manning/Task Organization

Subterranean threats require integrated capabilities and focused training. Israel has spent over a decade combating what some would argue is a strategic threat to their nation posed by subterranean terror tunnels built by Hamas and Hezbollah. To meet this threat head on, Israel created a specialized unit to take the lead on subterranean warfare. Within the Israeli Defense Force (IDF) Combat Engineering Corps, the Yahalom Unit is an engineering force that is organized specifically to address complex engineering problems in modern warfare.[25] The Yahalom Unit is made up of various elements that focus on specialized engineering tasks such as counter-terrorism, demolitions, EOD, maritime, and CBRN.[26] One subordinate element of the Yahalom Unit is The Samur (meaning "Weasel") Unit, which is specifically tasked with specializing in anti-tunnel warfare – detecting, clearing, and defeating underground systems.[27] The IDF have integrated a variety of skill sets and capabilities into The Samur Unit that have enabled them to achieve success in countering the threat posed by subterranean systems, while minimizing risk to its forces.

---

[25] Israel Defense Force, "This is the IDF's Plan to Combat Hamas Terror Tunnels," Israel Defense Force, accessed on 3 March 2018, https://www.idf.il/en/minisites/hamas/this-is-the-idf-s-plan-to-combat-hamas-terror-tunnels/
[26] *Ibid.*
[27] *Ibid.*

The United States does not necessarily need a carbon copy Samur unit, but the capabilities of such a unit are instructive of what task organized subterranean warfare teams should look like in order to meet the threat. Task organized subterranean warfare teams should be flexible enough to adjust to different operating environments, and should consist of combat engineers, demolitions experts, Explosive Ordnance Disposal (EOD) technicians, Military Working Dogs (MWD) teams, infantry, and must have robust robotic and autonomous systems capabilities. The goal with subterranean warfare teams is to have a level of proficiency, skill, and equipment that can be brought to bear to enable the clearance and destruction of subterranean threats without causing maneuver elements to lose tempo or initiative, and without placing friendly forces at excessive risk. Just as combat engineers and EOD use their specialized skills to reduce obstacles to enable infantry forces to penetrate, close with, and destroy enemy forces with speed, so too are subterranean warfare teams needed to deal with the complex threat.

E. Doctrine and Training

There is currently a gap in both doctrine and training for subterranean operations. If continued unchanged, there may be no experiences, lessons learned, Standard Operating Procedures (SOPs), or even a rough plan of action prior to facing an enemy utilizing underground environments in future combat. Warfighting publications, training and readiness manuals, and doctrine need to be updated and expanded. Most current publications devote less than five pages to subterranean operations.[28] These sections

---

[28] Publications referenced included: US Army Asymmetric Group Subterranean Operations Handbook (2015), US Army Tactics Techniques Procedures 3-21.50 Infantry Small-Unit Mountain Operations (2011), Marine Corps Warfighting Publication 3-17.4 Engineer Reconnaissance (2016), and Marine Corps Reference Publication 12-10B.1 Military Operations on Urbanized Terrain (2016).

provide mostly generic information and feature outdated tactics and capability sets. For example, most of the current military publications depict squad-sized elements establishing security, entering, and clearing subterranean systems with basic techniques and tactics that are simply above ground urban building clearance procedures.[29] These sections don't take into account any of the problems that make subterranean threats unique and more problematic than operating on the surface (see the paragraph above with the list of problems by warfighting function). As the military rewrites doctrine and warfighting publications to account for changes to modern warfare, subterranean operations must be rewritten in order to drive changes to task organization, training, requirements, and capability sets that will bring clarity to subterranean operations and enhance warfighting capability to face these threats in combat. Here again, the IDF provide the best foundation to start from. Through training, testing, and experience it can be molded to fit the needs of the United States military.

For training, the United States military should begin training its forces on subterranean operations during formal courses and as a part of standard unit training driven by requirements in training and readiness manuals. Subterranean training areas should be added to military range and training complexes to develop proficiency and familiarity. Additionally, subterranean threats should be included and evaluated during combat readiness evaluations and service level exercises. The subterranean environment is another environment, just like the urban, jungle, and mountain warfare environments. The only way to get proficient is to execute, test, evaluate, and then see what works, what does not work, and what requirements are needed to set conditions for success in combat.

---

[29] *Ibid.*

It does not need to be the top priority for the focus of unit training, but should an area forces are familiar with and have confidence with before facing a real threat down range.

F. Detection /Intelligence

The most critical aspect of defeating subterranean threats is detection. As with most military problems, there is no "silver bullet" solution to intelligence, or battlefield awareness. In order to successfully fight and defeat adversaries using subterranean systems, equipment sets with subterranean warfare-specific capabilities need to be developed and acquired. These systems need to be able to detect and map subterranean systems to a level of specificity that enable commanders to make decisions, rapidly address, and continue to drive offensive operations without ceding momentum, or the initiative to the enemy. Subterranean detection systems are broken up into two categories: above ground detection and underground detection.

Above ground detection methods are familiar to most military personnel and simply need to be adjusted to take subterranean threats into account. First, boots on the ground information collection has already been discussed in a previous section. While it is always an accurate way to confirm, or deny the presence of threats through physical presence and getting eyes and hands on, it should be a method of last resort and does not meet the intent of a having a capability that detects subterranean systems before placing friendly forces on the ground at relative disadvantage and high risk. The goal is to operate from a position of advantage in an environment where subterranean threats exist. Finding them when maneuver forces are already on top of them is too late. Second, a layered approach to standard intelligence collections assets provides great results – the more "ints" the better. HUMINT, SIGINT, IMINT, and MASINT can narrow the focus and reduce uncertainty for the location of subterranean systems. Third, manned and unmanned aviation platforms provide outstanding subterranean threat detection platforms. Most subterranean systems have surface cues that can identify the location of underground systems. Advanced aviation sensor packages can be used to identify such cues. Air and heating vents, generators, power lines, water pumps, construction equipment, and patterns of movement can be observed from the air, or a live video feed, to identify likely subterranean systems. There will always be ground cues for subterranean systems, though depending on the adversary and terrain, they might be very difficult to find, and might take a lot of time in addition to pulling high demand aerial platforms away from other combat tasks.

Below ground detection methods are capabilities that are still emerging for military purposes, but once realized can enable the United States to gain a significant advantage in fighting in a subterranean environment. First, ground-penetrating radar is a technology that uses electromagnetic waves (radio waves) to probe underground and can help identify underground systems by capturing and recording the energy that is reflected, scattered, or transmitted through the subsurface.[30] Ground penetrating radar is not perfect and does not always work well depending on soil composition and depth. Additionally, ground-penetrating radar requires significant time for data processing, which creates problems for its utilization in combat environments. However, as the technology improves and the equipment set becomes smaller and more manageable, ground-penetrating radar has potential as a force multiplier for subterranean warfare kits.[31] Second, seismic imaging is a field of technology that has been utilized by the

---

[30] Harry M. Joi, ed., *Ground Penetrating Radar: Theory and Applications*, (Slovenia: Elsevier, 2009), 4.
[31] Ibid., 145, 304.

private sector and geologists seeking to map, understand, and take advantage of the resources contained under the surface of the earth. Seismic imaging sends sound waves underground and records the echo with an array of sensitive geophones on the surface layer. After the data is collected and compiled, the technology provides a cross section of what the underground environment looks like.[32] Like ground penetrating radar, seismic imaging has yet to be converted into a viable technology for military purposes, but retains potential and continues to be improved by the private sector. Third, there are numerous newer technologies that are being developed and tested to identify and map underground threats. The Massachusetts Institute of Technology has been working with private firms to develop an unmanned aerial system with a sensor package that can survey large areas, and detect changes in magnetic fields to indicate subsurface systems.[33] The United States government has also been testing a vehicle mounted active seismic imager that impacts the ground, records the wave propagation, and can identify if there are openings under ground.[34] The system could be fixed to unmanned ground vehicles and employed forward of friendly forces. Additionally, private companies are experimenting with new systems that utilize Light Detection and Ranging (LIDAR) with autonomous systems to identify and map underground systems.[35] These imagery systems can provide highly

---

[32] The Chevron Corporation, "Seismic Imaging," The Chevron Corporation, accessed on 3 March 2018, https://www.chevron.com/stories/seismic-imaging

[33] Jon Harper, "Going Underground: The United States Government's Hunt for Enemy Tunnels," *National Defense Magazine*, accessed on 2 January 2018, http://www.nationaldefensemagazine.org/articles/2018/1/2/going-underground-the-us-governments-hunt-for-enemy-tunnels

[34] *Ibid.*

[35] Robert Gerbracht, (Marine Corps Service Fellow at Defense Advanced Research Projects Agency), discussion with author, 22 March 2018.

detailed three-dimensional images of underground systems that greatly benefit forces that must maneuver inside of, neutralize, or destroy these threats.[36]

There is no clear-cut technological asset that provides the military with an accurate, survivable, and reliable subterranean threat detection capability. However, there is traction in seeking a better solution. Israel has invested millions of dollars into research and development for their anti-tunneling research and development.[37] Israel is currently developing a 36-mile long sensor equipped underground wall along the Gaza border at a cost of $1.1 billion that is due to be complete in 2019.[38] The United States provided $40 million in military aid specifically to aid Israel for their anti-tunneling programs as well.[39] The DHS and US Army Research and Development Center have been actively testing and developing expeditionary subterranean detection equipment for those operating on the front lines against subterranean threats. The DHS has experimented and continues to test unmanned aircraft and ground vehicles equipped with radar technology that stream imagery of the subsurface terrain to agents in real time.[40] Additionally, numerous sensors, algorithms, and mapping programs have been developed seeking to image subsurface terrain in order to find tunnels and underground networks.[41] As previously mentioned, DARPA's Subterranean Challenge was unveiled in late 2017

---

[36] *Ibid.*
[37] Raphael S. Cohen, David E. Johnson, David E. Thaler, Brenna Allen, Elizabeth M. Bartels, James Cahill, Shira Efron, *From Cast Lead to Protective Edge: Lessons from Israel's Wars in Gaza*. RAND Corporation (Santa Monica, CA: RAND Corporation, 2017), 161-162.
[38] Dan Williams and Nidal al-Mughrabi, "Israel Says Foiled Hamas Bid to Rebuild Gaza Tunnel," Reuters, accessed on 18 March 2018, https://www.reuters.com/article/us-israel-palestinians-tunnel/israel-says-foiled-hamas-bid-to-rebuild-gaza-tunnel-idUSKCN1GU0A2
[39] *Ibid.*
[40] US Department of Homeland Security. *Tunnel Vision* (Washington, DC, 2009) Accessed on February 2, 2018, https://www.dhs.gov/science-and-technology/tunnel-vision
[41] Yiftah S. Shapir and Gal Perel, "Subterranean Warfare: A New-Old Challenge," The Institute for National Security Studies, Tel Aviv Univeristy (UNK), 54. http://www.inss.org.il/he/wp-content/uploads/sites/2/systemfiles/SystemFiles/Subterranean%20Warfare_%20A%20New-Old%20Challenge.pdf

challenging the scientific and engineering communities, and general public to develop breakthrough technologies to previously unimaginable capabilities for subterranean operations by 2021.[42]

Subterranean detection and mapping technology has been around for years, but has limitations for military employment, which include accuracy, size and weight, survivability, power, persistent communications, and in-stride navigation. Technology is a means to an end and not the be-all-end-all. However, technology for subterranean operations is a growing field and its continued development could lead to the establishment of critical tools that gains a significant advantage to the United States in a subterranean warfare environment.

G. Maneuver

Once an underground threat is detected and confirmed, it must be neutralized, or destroyed. Currently, military forces must either send ground forces into the tunnels to investigate, clear, and destroy tunnels, or use aviation delivered ordnance. Sending in ground forces plays directly into the operational objectives of weaker adversaries that seek to fight on a level playing field where they can inflict casualties on the stronger force, expand the length of conflict, create events that can be exploited in the information environment, and win by not being defeated. Maintaining a stand off approach and using aviation delivered fires has not been proven to be effective against the subterranean threat as damages can be easily remedied, or are unable to take out larger complexes leaving only part of the structures degraded. Subterranean threats clearly pose a problem to

---

[42] Defense Advanced Research Projects Agency, "DARPA Subterranean Challenge Aims to Revolutionize Underground Capability," Defense Advanced Research Projects Agency, accessed on 21 December 2017, https://www.darpa.mil/news-events/2017-12-21

ground maneuver and this problem cannot be fixed by precision fires. The solution to this problem is to utilize Robotics and Autonomous Systems (RAS) and Manned, Unmanned Teaming (MUM-T) to combat subterranean threats. Sending machines into subterranean environments to map, identify, confirm, and destroy underground complexes is preferable to sending ground forces down into a challenging, dangerous, and unknown environment.[43] As the United States military looks to incorporate technology and artificial intelligence, the subterranean environment is an area where human ground forces should be worked out of a job. RAS and MUM-T are areas of expanding military research, development, testing, and investment. Special Operations Forces, engineers, and EOD personnel have utilized RAS and MUM-T with great effectiveness for years, and it is currently all the rage in modern military discourse. The technologies are available, but must to be built with the specifications needed to effectively incorporate the assets for subterranean environments. The bottom line is that to thrive in and a subterranean threat environment, the United States must seek capabilities that avoid sending troops underground to maximum extend possible. If ground forces are required to physically operate in a subterranean system, then they must go underground armed with so much information and awareness that they know exactly what they are getting into before they go underground. This way ground troops can operate from a position of advantage and thrive in the environment just like they do above ground.

---

[43] Jon Harper, "Going Underground: The United States Government's Hunt for Enemy Tunnels," *National Defense Magazine*, accessed on 2 January 2018, http://www.nationaldefensemagazine.org/articles/2018/1/2/going-underground-the-us-governments-hunt-for-enemy-tunnels

VI. CONCLUSION

The United States needs to prepare to conduct subterranean warfare in future conflicts and must adjust concepts and capabilities to be able to thrive and defeat adversaries in this type of environment. The subterranean environment will be present on battlefields of the future. Just like the United States would find it unacceptable to send troops into a jungle environment without being familiar, or equipped to operate in the jungle, it is similarly unacceptable to look at historical examples and recent conflicts and not develop the capabilities to operate and thrive in and around subterranean systems. There are numerous goals that should be achieved. First, forces must be trained, equipped, and are prepared for subterranean threat environments before deploying. Second, once in theater, forces must be able to leverage standard collections assets and subterranean-specific equipment sets to enable rapid and accurate detection of subterranean threats, and three-dimensional mapping of the underground systems. These capabilities will in turn enable effective shaping of the subterranean environment, through destruction or neutralization, before committing ground forces to face the threats at close proximity. Third, task organized subterranean warfare teams must be developed with combat engineers, EOD technicians, demolitions experts, infantry personnel, and MWD teams that can detect, clear, neutralize, and destroy any remaining subterranean threats that haven't been shaped. These specialized forces will enable the primary maneuver forces to focus on aggression, initiative, and tempo against the enemy without being bogged down dealing with complex subterranean threats. Just as combat engineers provide the capability to breach large obstacles to enable infantry forces to rapidly penetrate and kill the enemy forces on the other side, so to are specialized subterranean warfare teams needed to achieve the same effect in modern complex terrain. Fourth, the United States must incorporate and expand technology, RAS, and MUM-T that is developed specifically for defeating subterranean threats. The subterranean environment is one where ground forces should be committed only as a last resort. RAS and MUM-T should be developed and incorporated to make modern day tunnel rats obsolete. The subterranean threat is not new and is not going to go away. What military forces do in between wars will determine whether or not it is an environment that leads to victory, or defeat.

BIBLIOGRAPHY

Cohen, Raphael S., and David E. Johnson, David E. Thaler, Brenna Allen, Elizabeth M. Bartels, James Cahill, Shira Efron. *From Cast Lead to Protective Edge: Lessons from Israel's Wars in Gaza.* Rand Corporation Monograph Series. Santa Monica, CA: RAND, 2017.

Defense Advanced Research Projects Agency. *DARPA Subterranean Challenge Aims to Revolutionize Underground Capability.* Defense Advanced Research Projects Agency, 2017. https://www.darpa.mil/news-events/2017-12-21

Dillon, Wayne. "Subterranean Warfare Considerations." Draft report, *US Marine Corps Tactics and Operations Group*, last modified September 1, 2015.

Drummond, Katie. "Lockheed Using Gravity to Spot Subterranean Threats." *Wired.* July, 15, 2010. https://www.wired.com/2010/07/lockheed-using-gravity-to-spot-subterranean-threats/

Estrin, Daniel. "Israel Speeds Up Underground Border Wall to Block Gaza Tunnels." *National Public Radio.* January 24, 2018. https://www.npr.org/sections/parallels/2018/01/24/579180146/israel-speeds-up-underground-border-wall-to-block-gaza-tunnels

Gentle, Gian and David E. Johnson, Lisa Suam-Manning, Raphael S. Cohen, Shara Williams, Carrie Lee, Michael Shurkin, Brenna Allen, Sarah Soliman, James L. Doty III. *Reimagining the Character of Urban Operations for the US Army: How the Past Can Inform the Present and Future.* Rand Corporation Monograph Series. Santa Monica, CA: RAND, 2017.

Ginsburg, Mitch. "How Hamas Dug Its Gaza 'Terror Tunnel,' and How the IDF Found It." *The Times of Israel.* October 16, 2013. https://www.timesofisrael.com/how-the-tunnels-in-gaza-are-dug-and-detected/

Hall, Benjamin. "Exclusive: Inside ISIS' Extensive Tunnel System." *Fox News Network*, October 23, 2016.

Harper, Jon. "Going Underground: The US Government's Hunt for Enemy Tunnels." *National Defense.* January 2, 2018. http://www.nationaldefensemagazine.org/articles/2018/1/2/going-underground-the-us-governments-hunt-for-enemy-tunnels

Headquarters US Army. *Engineer Reconnaissance*, ATTP 3-34.81/MCWP 3-17.4. Washington, DC: Headquarters US Army, March 1, 2016.

Headquarters US Army. *Infantry Small-Unit Mountain Operations*, ATTP 3-21.50. Washington, DC: Headquarters US Army, February 28, 2011.

Headquarters US Marine Corps. *Military Operations in Urbanized Terrain*, MCRP 12-10B.1. Washington, DC: Headquarters, US Marine Corps, May 2, 2016.

Helig, Donald M. "Subterranean Warfare: A Counter to U.S. Airpower." Master's thesis, Air Command and Staff College, 2000.

Heller, Or. "On the Way to the Tunnels." *Israel Defense*. July 14, 2016. http://www.israeldefense.co.il/en/content/way-tunnels

Howard, Courtney E. "Raytheon Develops Computer-Equipped Sensors to Locate Tunnels and Land Mines." *Military and Aerospace Electronics*. June 1, 2009. http://www.militaryaerospace.com/articles/print/volume-20/issue-6/news/news/raytheon-develops-computer-equipped-sensor-to-locate-tunnels-and-land-mines.html

Johnson, David E. *Hard Fighting: Israel in Lebanon and Gaza*. Rand Corporation Monograph Series. Santa Monica, Calif.: RAND, 2011.

Jol, Harry M. *Ground Penetrating Radar Theory and Applications*. Burlington: Elsevier, 2009.

Lambeth, Benjamin S. *Air Operations in Israel's War against Hezbollah: Learning from Lebanon and Getting It Right in Gaza*. Rand Corporation Monograph Series. Santa Monica, CA: RAND, 2011.

Lambeth, Benjamin S. *Air Power against Terror: America's Conduct of Operation Enduring Freedom*. Santa Monica, CA: RAND, 2005.

Lambeth, Benjamin S. "Israel's War in Gaza: A Paradigm of Effective Military Learning and Adaptation." *International Security* 37, no. 2 (2012): 81.

Lemothe, Dan and Carol Morello. "Securing North Korean Nuclear Sites Would Require Ground Invasion, Pentagon Says." *The Washington Post*. November 4, 2017.

Llopis, Jose L., Joseph B. Dunbar, Lillian D. Wakeley, Maureen K. Corcoran, Dwain K. Butler. "Tunnel Detection Along the Southwest US Border." April 1, 2016. https://www.researchgate.net/publication/269122889

Macartney, John D. "John, How Should We Explain MASINT?" In *Intelligence and the National Security Strategist: Enduring Issues and Challenges*, edited by Roger Z. George and Robert D. Kline, 169-179. Lanham, MD: Rowman and Littlefield Publishers, 2006.

Magnuson, Stew. "Holding the Low Ground: Daunting Challenges Face Those Waging Subterranean Warfare." *National Defense* 91, no. 639 (February 2007): 20-22.

Medalia, Jonathan. "Nuclear Earth Penetrator Weapons." *Congressional Research Service* Report for Congress. Washington, DC: 2003.

Nguyen, Kha M. "Learning How to Mow Grass: IDF Adaptations to Hybrid Threats." Master's thesis, School of Advanced Military Studies, US Army Command and General Staff College, 2017.

NoCamels Team. "7 Things We Know About Israel's Secretive Anti-Tunnel Tech System." *NoCamels Israeli Innovation News*. November 1, 2017. http://nocamels.com/2017/11/israel-anti-tunnel-technology-hammas/

Pappalardo, Joe. "High-Tech Border Patrol: 5 New Tricks to Find Smuggler Tunnels." *Popular Mechanics*. September 30, 2009. https://www.popularmechanics.com/military/a2487/4244235/

Pappé, Ian. *Gaza in Crisis: Reflections on Israel's War against the Palestinians*. Chicago, IL: Haymarket Books, 2013.

Ryan, Missy. "Israeli Offical Bets Advances in Anti-Tunnel Technology Will Secure Gaza Border." *The Washington Post*. March 6, 2018. https://www.washingtonpost.com/news/checkpoint/wp/2018/03/06/israeli-official-bets-advances-in-anti-tunnel-technology-will-secure-gaza-border/?noredirect=on&utm_term=.55783e1e358d

Shapir, Yiftah S., and Gal Perel, "Subterranean Warfare: A New-Old Challenge," *The Institute for National Security Studies at Tel Aviv Univeristy*, 53, http://www.inss.org.il/he/wpcontent/uploads/sites/2/systemfiles/SystemFiles/Subterranean%20Warfare_%20A%20New-Old%20Challenge.pdf

Underground Facility Analysis Center. "UFAC Digs Deep to Find Covert Facilities." *Communique*, May/June 2010, 29.

US Army. Asymmetric Warfare Group. *Subterranean Operations Handbook V3*. Fort Meade, VA: US Army Assymetric Warfare Group, 2015.

US Army Training Circular 3-21.50. *Small Unit Training In Subterranean Environments*. Washington, DC: US Army, 2017.

US Department of Homeland Security. *Tunnel Vision*. Washington, DC, 2009. https://www.dhs.gov/science-and-technology/tunnel-vision

Vergun, David. "Tunnel Detectors Ferret Out Enemy Below Ground." *US Department of Defense Army News Service*. June 23, 2017. https://www.defense.gov/News/Article/Article/1226944/tunnel-detectors-ferret-out-enemy/

Watkins, Nicole, and Alena James. "Digging into Israel: The Sophisticated Tunneling Network of Hamas." *Journal of Strategic Security* 9, no. 1 (2016): 84-103.

Weaver, Mary Ann. "Lost at Tora Bora." *The New York Times*. September 11, 2005.

www.ingramcontent.com/pod-product-compliance
Lightning Source LLC
LaVergne TN
LVHW061343060426
835512LV00016B/2654